TOMMY PAGÁN

STOP PUTTING LIPSTICK ON A PIG

THE FUTURE IS EXPERIENCE-LED AND HUMAN-CENTERED

BRIGHTRAY
PUBLISHING®

We help busy professionals write and publish their stories
to distinguish themselves and their brands.

(407) 287-5700 | Winter Park, FL
info@BrightRay.com | www.BrightRay.com

ISBN: 978-1-956464-45-0

Published in the United States of America.
BrightRay Publishing ® 2023

This book is dedicated to my family, Nilma, Gustavo, and Penelope.

To Mom and Dad:

Thank you for your thoughtful love, time, and support, along with the many life lessons you imparted to me. You taught me the importance of education and work, to dream, be respectful, and, most important of all, family. I wish you lived a little bit closer, but I hope we can see each other more and enjoy life together.

To my sisters and my brother, Madelyn, Patricia, and Kevin:

Thank you for an incredible life well lived and for always being there for me, my family, and each other. Thank you for the wonderful adventures, memories, laughs, and love throughout our journey together!

And to all of the designers out there who are striving to make a difference in the world:

Thank you for your passion and dedication. We need more people like you in this world.

TABLE OF CONTENTS

INTRODUCTION

> *To design is to devise courses of action aimed at changing existing situations into preferred ones.*
>
> *—Herbert Simon*[1]

Have you ever encountered an app that was so hard to use that you couldn't even figure out how to sign up for an account? Or a website that constantly logged you out?

What about those kiosks at the airport? Would you rather talk to someone behind the counter? The line is so long . . . Could they just hurry up? Oh no, where'd your toddler run off to? Where is the back button on this thing? Now, take a deep breath; it's time for the TSA . . .

All of these are examples of common pain points that create frustrated users, and frustrated users have bad experiences. These pain points are found at all interaction levels. Some are only surface-level. Some are at the human-computer interaction. Some are structural. Some are more complex strategic problems. A frustrating experience, whether it be with a brand, product, or service, is a sign of a design flaw.

When things don't work like you would expect them to, you feel personally inconvenienced. Your journey navigating a website or checking in at the airport is brought to an abrupt halt. More often than not, these poor experiences are the product of business teams, developers, or engineers working without a designer's involvement. A designer's job includes making sure any discomfort with a product or service is minimal but, sometimes, executive decisions ignore the design work. Choices end up being made without research or analysis, and the people who suffer are the users.

As a user experience designer (UX), I've worked with companies across all industries and have encountered the same problem over and over again: there continue to be misunderstandings about the role of a designer and the benefits of a well-thought user experience. Leadership must learn to incorporate design into all levels of their businesses, from products to engineering teams, and understand, once and for all, that design goes beyond surface-level problems.

Despite the plethora of articles, books, and reliable sources of information on user experience and the benefits of a user-centered approach, designers often get confused with decorators. Stakeholders only perceive value if their designers produce visual graphics or make their products "pretty." I had started to give up on correcting these people but, instead, I decided to write a book.

The truth of the matter is: if you are using your design team purely for beautification, you are only scratching the surface of their capabilities. If you think they are happy working like this, you are wrong. Luckily, now is the best time to bring change to your UX department or product team.

In his book *The Sciences of the Artificial*, Herbert Simon defines design as "devising courses of action aimed at changing existing situations into preferred ones."[1] User experience includes all of the interactions a user has with your product or service. Nothing is set in stone; it can always be improved.

Much of my early experience in UX as a designer was in the healthcare industry, building user interfaces for electronic health records in Puerto Rico. Our teams were incredibly collaborative and progressive. We had not only project managers, software engineers, and architects, but also nurses, doctors, and healthcare professionals all in the same room. There was an interest in discovery and exploration from all team members. It was an incredibly rewarding experience that raised my expectations and showed me how I wish more teams would work today.

Unfortunately, this did not last long. The company expanded and leadership changed. They brought some experts to help us improve our efficiency; I was moved to a cubicle and asked to estimate my hours on a spreadsheet. There were no more discovery sessions or prototyping. It was build, build, build, then make it look pretty.

After five years of working on web applications, I shifted to marketing websites. One of those first jobs had a gaming platform. One of the company's leaders had been a heavy user of the platform since he was 14. He constantly made suggestions via the forums, demanded new features, and was eventually given a job.

At the young age of 21, he was vice president. He would assign tickets via the project management app. These were often very vague, and I would come back to him with the usual questions to understand the actual task and formulate an actual design brief (as I had been properly educated to do.) One of those times, he snapped at me, "Why do you ask so many damn questions? Why don't you do your job?"

"That's precisely what I'm doing," I said, "my job." Needless to say, I moved on from that company.

Blindly designing makes no sense, and treating your team with hostility is no way to work. As I interacted with more companies over the years, I found that design teams were being given pixel-level direction—essentially micromanaging each decision—by executives and managers with no design experience whatsoever. This leads to more frustration and miscommunications. With this lack of design understanding, design teams are left to tackle problems on a surface level, and are tasked with putting lipstick on a pig: covering up problems with aesthetics rather than solving problems at their core. Unfortunately, this is the sad situation in many companies that I have worked for.

After almost a decade of searching for meaningful design work, my luck changed when I became the lead UX designer for a consulting company. Instead of micromanaging every idea and input, my manager's refreshingly trustful approach allowed me to thrive, learn, and explore. I was more creative and productive without hesitation or fear.

"You're the expert," she said. With her unwavering support for almost everything we designed, it was an incredibly liberating experience.

Support like this for designers is rare, especially when many businesses hire you as a contractor instead of a consultant. Team members are seen as interchangeable, and people are hired and fired on whims. The culture around user experience needs to shift, and this will only happen from the top down.

Everyone can benefit from this mindset evolution. Not every company needs to be a huge tech conglomerate to benefit from a well-oiled UX design team. There are small to mid-sized businesses that just need the right leadership and attitude to actually capitalize on what UX design and a designer mindset can offer.

I have a constant desire to learn and understand the world around me. I'm not afraid to stand on my own and let others know if I think something is wrong or irrational. This, in large part, is why I've written this book.

Real change is only possible with honest efforts in empathy towards your users, humility to learn, and trust in your teams. It takes patience and understanding for your teams to create an environment where they can truly delight your user and customers. Then, and only then, will you finally be able to put the lipstick down.

CHAPTER 1

Start with the Experience: The Basics of UX

> *You've got to start with the customer experience and work backward to the technology . . .*
>
> —*Steve Jobs*[1]

It may seem obvious how much we interact with technology in the span of our lives but think about how much you interact with technology in the span of a day. An hour. A minute.

Your experience is completely unique to you and no one else. Even so, each computer, phone, website, or app you engage with has been designed so people around the world can use it. You are a "user" any time you interact with a service or product. User experience includes all of the users' emotions, beliefs, preferences, perceptions, physical and psychological responses, behaviors, and accomplishments that occur before, during, and after use.

Much of our experiences with services and products are digital, and their development is often technically complicated. That doesn't define what a UX designer's job is focused on though. As design expert Don Norman explains, the experience of a user goes beyond just interacting with the product.[2] You could be telling a friend about your time with the product, and even that would be included in a user experience.

UX design is a process that focuses on understanding users and their goals, behaviors, and context to improve user satisfaction with a product or service. This involves the design of the entire process of acquiring and integrating the product, including aspects of branding, aesthetics, usability, and function. If you've hired UX designers, they should be taking into account how your customers interact with your business on all levels.

By having leadership, not just the designers, thinking about customers' and users' experiences, any organization can gain tangible benefits from UX design. It can go as far as defining a business strategy. User testing and research can help save money by identifying problems before they happen. Risks can be reduced with a well-researched implementation process. User adoption will increase by providing the products and solutions they want. Improved efficiencies for organizations and their customers lead to happier employees. It just requires patience and planning.

Focusing on that experience for users and customers does pay off. One study by Forrester found that companies that prioritize the customer experience generate 2.5 times more revenue than those that don't.[3] Another study found that 86 percent of customers are willing to pay more for a better customer experience.[4] Proper UX design is the key to creating delightful experiences for your users. Business owners and managers, however, need to understand what UX is and what it is not.

The main distinction that needs to be taught is the difference between user experience (UX) and user interface (UI). I am amazed at the confusion that still dominates these topics. Oftentimes, people confuse the two terms and refer to them as UX/UI as if they mean the same and are interchangeable. They are not. The user experience does not start or end with a computer screen nor does the user interface. The UI is the "bridge" between human and computer interaction. It could be a screen, a mouse, or a keyboard. Nowadays, with devices using digital assistants such as Alexa and Siri, natural language is the UI, often referred to as voice user interfaces (VUI). Natural gestures are another form of UI, often used with touch screens and touchless interfaces in the gaming and healthcare industry.

As you can see, UIs are much bigger than the colors, buttons, and typography you see. Because the UX is so intangible, it becomes difficult for non-designers to explain it in words, so describing the UI is often how they define the user experience.

There are ways the UX can be defined and measured. Data and research is analyzed against the intended users' expectations and used to determine whether the design meets, exceeds, or completely underwhelms them. One of the ways that designers hone in on a user's experience is through personas. Personas serve as models for potential users or buyers, like fictional characters, to keep real users in mind during the design process. Designers have to be careful not to focus too much on personas as they are useful tools for guidance rather than constraints to the design process.

You can't forget: users are more than customers. There is a school of thought that finds it offensive to call people "users" and prefers CX (customer experience) instead. Initially, the term "user experience" was meant to include all interactions with the company, and its products or services. Since then, the term "customer experience" has become what the original term was meant to be. The truth is that everybody who uses your products or services is, in fact, a user. But not everybody is a customer. A customer can buy a product from your company but not use it. So, what keeps a customer using what they've bought?

Overall, it's a positive experience with the product that brings something unique and innovative to the table. The best way to break through the status quo is through the use of "design thinking." As Tim Brown of IDEO puts it, design thinking is a human-centered method of observation, collaboration, fast learning, visualization, rapid prototyping, and business analysis to influence innovation and business strategy.[5]

Design thinking helps businesses achieve a balance between what is **viable** for business success and what is technologically **feasible** while considering the real needs and desires of people. Thinking like a designer can help organizations transform the way they develop products, services, and strategies.

This process consists of three major phases of activities. The first is Inspiration, which requires designers to research their users and understand their wants and needs. The second phase is Ideation. This lets the designer loose creatively, allowing them to come up with as many crazy concepts and ideas as possible. Then, the last phase is Implementation. Here, rapid prototypes are built for these ideas, creating tactile products for feedback. Testing allows users to work with these prototypes and see if goals have been met.

Even when goals are met, this process is ongoing. Products and services can be continually improved by leading with the experience or by incorporating a simple user-centered design principle into your organization. This means making sure that every single decision you make as a business is driven by the needs of your users and customers. It means putting yourself in their shoes and understanding their wants, needs, and pain points. Listening sometimes is just not enough.

At one company, I had a boss who insisted on listening in on client meetings and oftentimes would attempt to lead them. He thought that he was "doing UX design" by talking to users, but he didn't know better.

We often asked our users to walk us through their process of completing a task, intending to understand any inefficiencies, and then make any suggestions or proposed solutions. During one of those meetings, my boss witnessed a client take 15 minutes to go through 17 different screens just to log in, struggle to keep track of what she had just done in a spreadsheet and grow overwhelmed by all of the moving screens. Even still, my boss said, "Well guys, they can do it, so we don't have to spend any more time on this." He was implying we didn't have to implement this feature in our new iteration of the product because the users can always go to the legacy product and do the task. My colleagues and I were shocked by this, and we exchanged emojis with eyes popping out along with other jaw-dropping animated gifs.

Proper UX design would take all of our observations of this interaction into account. It took this user a considerable amount of time to do a simple task—at least 15 minutes! If you multiply how much time it took by how many times it needed to be done a day and then by how many users would have this same problem, the users' patience, and goodwill are wasted on just logging in. I don't know if my boss didn't see her struggle or just didn't want to, but his indifference meant the product suffered. That is why listening to your users' needs is not enough when you have the wrong people with the wrong goals listening to users. It's not "extra" work if it's necessary to make a better product. The goal is to make sure that the users have a positive experience with your product or service at every single touchpoint.

Creating positive experiences at every touchpoint may seem like a cost-inhibitive process, but there are ways to ensure efficiency in the development process. There are two primary methodologies for development: lean and agile. Many businesses like to capitalize on one or the other, but this is limiting. Each one has its own strengths and weaknesses and should be used to complement the other. The key to their success is tackling development and production in iterative steps.

Originally, the waterfall model was popular in early software engineering as work flowed in a linear, sequential path down. Designers only moved from one phase of development to the next once the prior phase was completed. This process limits flexibility and often leads to redesign, redevelopment, and ultimately higher costs. As the limits of the waterfall methodology were realized, designers began to explore other options, like lean and agile.

Lean as a management tool was developed by Japanese carmaker Toyota's production line and popularized by Jim Womack and Dan Jones in the 1990s.[6] It encourages quick delivery but focuses more on the value for the user or customer, reducing any waste in the development and production process. If relied on too heavily, the necessary cost reductions to maximize value are often abused by managers, resulting in underpaid employees.

Agile's roots are in software development, where software is delivered as quickly as possible once it is working.

Instead of large quantities of software being put in front of customers at once, the frequent deployment allows quick turnaround on feedback. This feedback influences the work in the moment, no matter what step designers are in the development process. It is a cross-functional effort by teams focused on one iteration at a time. When applied to large companies, it can often be inefficient, and if improperly used, it overemphasizes the process over the product.

The best approach to the user experience design process is to combine design thinking, lean, and agile together. Design thinking creates a systematic approach to identifying users' problems. Lean development allows the right product to be built in the most efficient way possible, integrating feedback into the development process. Agile ensures the product is built right, providing incremental innovation towards the solutions for the users. More often than not, the solution to the problem is not an all-or-nothing response.

Still, the most important part of the whole process is empathizing with the user. A humble and understanding mindset is necessary for real problem-solving and continuous improvement. This human-centered approach is what brings true value to the users and ultimately to your company. As writer Robert Lee Read said, "Creating products that effectively meet the needs of the customers is more important than doing the wrong thing energetically."[7]

If you're not putting the user and customer experience first, then what are you putting first? The technology? The features? The money? Your own ego? When you don't prioritize the user and customer experience, it shows. It shows in the design of your product, it shows in the way your employees interact with customers, and it most definitely shows in your bottom line.

CHAPTER 2

Human-Centered Design

> *Human-centered design. Meeting people where they are and really taking their needs and feedback into account. When you let people participate in the design process, you find that they often have ingenious ideas about what would really help them. And it's not a one-time thing; it's an iterative process.*
>
> *—Melinda Gates*[1]

Human-centered design is a mindset, vision, and goal for any organization looking to create a product or service that will be used by actual human beings. This goes beyond just the "humans" using these products and extends to all the employees and stakeholders involved in the production process. Humans are making these products for other humans, and everyone should be taken into account.

It may seem like a no-brainer to have people in mind when designing but, more often than not, humanity falls by the wayside. It's not a new or revolutionary way of thinking either. Human-centered design can be traced all the way back to Plato's Republic. Plato believed in seeking advice from all members of a community when designing a fair city-state. What can be classified as "grassroots democracy" is really participatory design and collaboration toward building the best product possible.[2]

Humanizing the design process brings the stakeholders, designers, and users together under an umbrella of empathy. This humanization can only be implemented from the top down though; heads of organizations need to embrace this mindset, or they break down. How can a designer empathize with a user if they are building a product to the specifications set forth by a CFO? Internal alignment is necessary so that good customer experiences can be properly designed.

It is necessary to remember that designers are solving problems for other human beings using their product and not for the "business." This doesn't mean that a business sensibility is unnecessary in the design process, but the real value comes from solving challenges at the end-user or customer level. To optimize that value, real core issues need to be solved—and not just the symptoms. Nobody wants to use a product that feels like putting a band-aid over a bigger problem.

To create an environment where human-centric design can actually flourish, leadership needs to commit to a bold human-centric strategy. By focusing on humans, you include your employees, your users, and your customers as well as their microcosms, communities, cities, regions, and nations. If the effort is disingenuous (as I have experienced many times), then your business will not actually reap the benefits of a human-centric strategy.

One of the ways to understand an earnest implementation of human-centric design is by looking at everything as a system.

Every part of your company, from the customer service department to the C-level executives all the way to the warehouse distributor, is part of an interconnected system. This is classified as "systems thinking," meaning every problem within this system is interconnected.

By designing seamless experiences across service, physical, and digital products, a holistic system is built with the ability to actually diagnose and solve problems. Donella Meadows puts it best: "A system is a set of related components that work together in a particular environment to perform whatever functions are required to achieve the system's objective."[3] This recognition of interconnectedness expands the choices available to you and creates more satisfying, long-term solutions.[4] When humanity and empathy are ignored, you end up with low adoption rates, low engagement from both team members and users, and plenty of features with no real value to them and, therefore, the business.

Businesses have tended towards customer-centric design, rather than human-centric design, to enforce that interconnectedness. Customer-centric thinking leaves a large part of your business or organization out of the equation. Stakeholders will point to increases in revenue, but that is only a small snapshot of the whole picture.

McKinsey research reveals that improving the customer experience has increased sales revenues by two to seven percent and profitability by one to two percent. In addition to this, the overall shareholder return has increased by seven to ten percent.[5] In theory, these numbers are enticing, but true customer focus leaves the employees out of the picture.

An example of this can be found in Amazon's business model. They can be classified as almost completely customer-centric yet lacking in human-centric vision. To be able to receive your purchase in less than two days is a modern marvel, but who really pays that price? More often than not, it's Amazon's employees.

Jeff Bezos has long felt that hourly employees who spent years working for the same company were inefficiencies that needed to be eliminated. His solution has been to replace workers before their productivity decreased, and he was forced to increase their pay. The company even offers bonuses to people who are willing to quit after a certain point. Turnover at Amazon is higher than many other companies with an annual rate of about 150% for warehouse workers.[6]

This dehumanization of employees continues in how warehouse workers are treated in their day-to-day operations. Bathroom breaks are reduced to employees relieving themselves in water bottles for fear of being less productive. Safety has been skirted, even though Amazon has pledged to become "Earth's safest place to work." However, that didn't stop injuries at its US warehouses from increasing between 2020 and 2021.[7]

The employees have paid the price for an "ideal" customer-centric model, but it is no longer sustainable. Leaked internal research shows that Amazon could run out of people to hire in its US warehouses by 2024.[8]

Richard Branson's approach to management and corporate philosophy, on the other hand, reflects a more human-centric design: "Clients do not come first. Employees come first. If you take care of your employees, they will take care of the clients."[9] Branson's Virgin group puts its employees and customers ahead of its shareholders because they understand the connection between accepting empathy and good business.

The ability to connect with users is not inherent and requires insight in the form of research. Research informs every decision made during the design process. As designer Erika Hall has said, "Design happens in context. And research is simply understanding that context. Research also happens in a context as well. Your professional environment should inform the research activities you choose and how you go about them."[9] If that environment is unhealthy and not human-centric, that affects all the data collected, as well as its analysis.

That analysis should be unbiased and free of any opinions from stakeholders. Problems should be dealt with at the source within the system rather than solely addressing symptoms. This root cause analysis highlights the system's vulnerabilities rather than individual human errors and less actionable root causes. Research is often overlooked by management and stakeholders as it is seen as a superfluous step in problem-solving. In reality, it is a human-centric necessity to find what is plaguing users and designers.

One of the best places to gather insight is with the customer service department of your organization. Customer service is where individual human needs and expectations crash headfirst into reality. Again, management tends to overlook customer service staff when they have some of the most vital interactions with users directly. The research necessary for human-centric design requires empathy and humanity, which are often overlooked for more "business-sensible" decision-making.

Leadership should know that most business-sensible decisions are influenced by human-driven data. Data should not be feared by stakeholders even if they are not knowledgeable in interpreting it themselves. That fear can hold back a business from internal innovation and potential automation.

The way Amazon uses data is probably influenced by that desire for high turnover, using humans as mere cogs in the machine. Data doesn't need to make employees obsolete, and it can actually be used to retrain employees in more efficient ways. Work should be kept meaningful instead of tedious and mindless. Employees will be trusted to take care of the customers, as long as management keeps their humanity in mind.

Customer experiences can be innovated as well, with research shining a light on the predictive capabilities of systems. By looking forward, designers can anticipate customer journeys in more empathetic ways, creating consistently delightful experiences.

Users, employees, and designers are more than just a number to be tallied by the business. Yes, the research may reduce everyone to a number, but those numbers need to be applied to exciting, innovative design. This can only happen when management, stakeholders, and executives make institutional changes from the top down. The way businesses look at "human resources" is a sign all on its own. Humans are not resources; they are humans. The department should be called "business resources" because it handles the inner workings of the business. In almost 20 years, I have never experienced a "human" resources department that works for or on behalf of me, though I did encounter a tiny exception once while working at the IT solutions company, Mindtree.

The dehumanization of employees and users only makes the product and services suffer. That dehumanization may not be management's intention but all it takes is a few incompetent, insecure, or non-collaborative leaders to create a system that doesn't have the human experience in mind.

CHAPTER 3

Horses, Hippos, and Other Wild Animals

> *If I had asked people what they wanted, they would have said faster horses.*
>
> *—attributed to Henry Ford*[1]

"Shots fired, shots fired!" exclaimed a colleague of mine, having read a post written on LinkedIn by an executive in our company aimed at another executive in the same company—our boss. My colleague knew that I would be interested in the topic because not too long ago, I had shared with him a graphic titled "The Dangerous Animals of Product Management." In the post, the executive referred to our boss as the HiPPO in the room.

There are many kinds of "animals" in companies who not only hinder innovation, creativity, and productivity but also bully others into submission, affecting the performance of the team members and the outcomes. These animals have been used to represent individuals, mindsets, or entire cultures in the workplace for the past decade or so. Please keep your eyes open so you can recognize them and do something about them.

Too often, people think of Henry Ford as the automobile mogul who revolutionized assembly line mass production, but before his rise, he was an engineer working for Thomas Edison. A famously misquoted saying proposes that if Ford had asked people for their solutions, they would have wished for faster horses.[1]

The problem wasn't slow horses; it was the physical limitations of animal-propelled transportation. You can't start with a solution if you don't understand the problem. And sometimes, the people who give you solutions are too close to the problem to see the whole picture. Suddenly, leadership becomes a roadblock to actual design work by suggesting too many solutions without addressing the problem.

Too often, upper-level management feels the pressure to have all the answers all the time, so when the design process doesn't give them the answers they want, they start providing ideas for teams to execute anyway. UX design doesn't start with answers—it starts by looking for problems. That stakeholder who never studied UX but has the most sway in the room is usually a HiPPO.

HiPPO — Highest Paid Person's Opinion

HiPPOs are the worst animals to have on your product team not because they have an opinion, but because they tend to listen only to their own and shut out everyone else's. Once the HiPPO in the room speaks, the conversation usually goes two ways.

The group either agrees with the HiPPO's opinion (echoing it), or the group keeps its thoughts and opinions to itself due to fear or conformity. They are the "know-it-alls" whose factless opinions are the only ones that matter. It's irrelevant to them if you are the expert in the area or the one who was actually hired for that particular job. HiPPOs see themselves as superior, smarter, and exceeding your education and experience in the content matter. No data can make the HiPPOs change their mind. They are there with one agenda only: "My way or the highway." Who dares speak against the mighty HiPPO? Who takes a stand against them? Most definitely no one for fear of retaliation or being terminated.

Though HiPPOs are the "highest paid persons" in the room, they are not solely found in C-level executives. HiPPOs can be encountered at any level in the organization. Like real-life hippos, they have little ears and big mouths. Also, if they feel threatened or you stand between them and "the water they live in," they will charge at you. So watch out! Their charge can be fast and shut down valuable contributions to your products.

I understand that sometimes situations are delicate: there might be decisions with million-dollar effects, and that's a lot of pressure. Leadership is not expected to always have every answer, but this is exactly why they should defer to the experts who they hired. Gut feelings do not drive innovation or solutions: diversity, teamwork, and exploration do.

If you're reading this and you don't think you have ever worked with a HiPPO before, take a moment to look at yourself and ask, "Could *I* be the HiPPO?" Everyone's guilty of it at some point or another, but by admitting it, you can put an end to it. Try asking for feedback from team members before giving your opinion. Trust your team and encourage them to speak their mind without retaliation. Until something is done, it's bad for team morale, and it's hurting your profit. It's not only HiPPOs you have to look for in the team, though they tend to be quite common.

ZEBRA — Zero Evidence But Really Arrogant

Remember the executive accusing our boss of being a HiPPO? Well, she is not without blame here. Meet the ZEBRA. They are experts in everything, even though they aren't. Being overly confident in their own intuitions leads a ZEBRA down an opinionated path just like the HiPPO except they don't need the highest salary to be heard. Their cockiness is enough to sway decisions. The product will suffer if a ZEBRA's intuition replaces research and evidence. Unvalidated products or features get built as the project progresses in the wrong direction.

In the case of the ZEBRA from our example, a name had to be picked for a product, and leadership had some ideas but felt they were weak and/or outdated. The ZEBRA executive, who had a very advanced degree, was assigned the task to come up with a name for the product. Three names were chosen and focus tested. The people, including executives, voted on this name. When the ZEBRA finally presented the name to the larger company, we, the product team, did not hold back. The pushback from us was unanimous. These names were even weaker than the ones previously chosen by the other executives. I mean, these names were laughable and borderline inappropriate for the category of product at best. So, it was *too easy* for the executive team to choose one of the names they initially came up with. That is when the ZEBRA wrote that post about the HiPPO.

If uninformed managers are holding onto a cognitive bias, the Dunning-Kruger effect rears its head. The Dunning-Kruger effect occurs when people with low experience in an area of knowledge overestimate their own abilities and in this case, create products and services disconnected from humanity. Ignorance just compounds an already difficult process.

Even if they are an expert in a field, someone can be a ZEBRA. Sometimes that expertise is what blinds them; they think they don't need to learn more about the work. I admit to being a ZEBRA at times, given my natural tendency to rely heavily on my intuition. But in this case, my training as a designer and the design thinking process is what prevented me from going too far down that road throughout my career. No matter what, you need evidence. Embrace data; stay humble and curious.

RHiNO — Really High-value New Opportunity

If you see a RHiNO charging your way, they probably have requests for new features. Their goal is often to bring in new clients with a trendy design, or maybe there is a potential deal if the product "just had this one feature." In reality, these features most likely aren't necessary.

More often than not, the RHiNO is from sales. I have nothing against salespeople; they have their own complicated tasks to carry out. I have partnered with them on multiple occasions as they are often closest to the people we are building and designing products for. However, more often than they would like to admit, they're either reacting to symptoms and ignoring the problem or taking orders directly from the customer. If these requests are not properly filtered, it could detract from the best possible product for the user, all to focus on what will convert someone into a customer.

The RHiNO is nearsighted in this way. They are worried about sales, which affects their quotas, and could be called out by their manager. It's their fear (a fear of the boss or a fear of missing out) rather than a strong vision that guides them. The RHiNO should look further out to see the opportunities that come from creating delightful experiences. Be very careful, and always weigh those features against the bigger picture and the long-term value of the product. When you add many features without proper testing, it could attract another type of animal.

WOLF — Working on Latest Fire

The WOLF has their attention divided by trying to put out all the different fires that arise in the day-to-day. Without their focus, a lack of cohesion in the work and product forms. Rather than figuring out what is causing the fire, the WOLF rushes to the flames, often pulling in parts of the team to help them.

When everything is important, then nothing is. They move from problem to problem too quickly, and they may even cut corners to meet deadlines or run to another fire. With that overly agile mindset, a WOLF rushes the product: designers are forced to expedite designs, and developers implement watered-down versions of features. On top of that, the proper QA was skipped, and now it's in production. The proper checks and balances are in place for a reason—to deliver something functional. If a manager rushes a project, the product will be badly built and buggy. They will be unhappy and blame the product team when, in reality, it was because of their WOLF tendencies.

The WOLF needs to feel "needed." It gives them a sense of purpose, of identity. They often cut corners because if they deliver something quickly, even if it's not workable, they are valued by the management team.

The main problem with the WOLF is the delivery of a substandard product. This causes an accumulation of design and technical debt. Technical debt, coined by Ward Cunningham, is the additional costs you pay after you launch and release software too quickly in an effort to cut corners.[2] In the end, it all catches up to you, and the business pays the price.

Design debt stems from all of the important design concepts or solutions that were skipped in order to reach short-term goals. In the moment, it may not seem like that big of a deal as long as the product works. Those small hiccups and bad, little experiences build up in the user's subconscious; it seems unprofessional. Mark Boulton calls this the marginal degradation of brands.[3] If your brand's image suffers from a poor product, it's hard to win the user back.

A WOLF may feel like they saved the day by jumping from fire to fire, but they actually pulled designers away from work, slowly but surely accruing debt. Without that focus, leadership leads designers in the wrong direction, and the product suffers. "Good enough" is the mother of mediocrity.

SEAGULL — Seagull Manager

A SEAGULL manager is often removed from your day-to-day work, flying high above the rest of the teams. When they feel like it, they swoop in, cause a mess, and then fly away, just like a seagull.

They might have good intentions but, because they are so removed from the current situation and flying so high up, they are out of touch with what is really going on. Their efforts to come in and "save the day" are similar to a WOLF, but their lack of context complicates the dynamic. It even takes away from the design team's agency, especially if the SEAGULL doesn't always have the best intentions.

Sometimes, a SEAGULL is a real "negative Nancy," and by being so removed from their designers and the project at hand, it makes it easier to be pessimistic about the work. They aren't as invested and don't see the long-term consequences of their "help," so when things don't go as planned, a SEAGULL usually isn't around. When they swoop in, all they see are problems to "poop" on.

HONEY BADGER — Honey Badger Manager

Then, there is the HONEY BADGER. If you have ever seen that famous YouTube video, you know that the honey badger does not care. The honey badger does not give a %$@!

These team members and managers are only in it for the money. They are not invested in the work. They don't have the users' best interest or their team members in mind. If they hold key positions, they can compromise the quality of the whole product.

The HONEY BADGERS do not learn from their mistakes; they pretend to listen and nod with their heads in approval, but in reality, they can't wait for you to shut up. If they are managers, they do not provide feedback to their teams or care what users have to say. They only do the bare minimum just to ensure their survival.

With all of this wildlife, what causes these animals to attack their own teams? A lack of trust lies at the heart of the problem. If a HiPPO trusted the team's voice over their own if a WOLF trusted the team to put out the small fires if the RHiNO trusted the team's work to create an ideal product if the HONEY BADGERS would only give %$@! then these animals could focus on the more meaningful tasks of management. A large part of these stakeholders' lack of trust comes down to unrealistic expectations.

You need to have a *human-centered vision.*

Knowing what your designers are capable of seems like a basic aspect of UX leadership but isn't possible without a willingness to learn, humility, and empathy. If you hire a UX designer, and you don't obtain the results you expected, you will feel betrayed. You may even begin trusting yourself over anyone else. A human-centric mindset, however, can clear any roadblocks for you and your team.

Human-centric design requires humility. Without humility, engineers, managers, and designers will disregard user feedback, blaming each other or the user instead of potentially bad design. Pointing fingers doesn't solve problems; fixing the root cause does.

Designer Don Norman points this out in his book, *The Design of Everyday Things:*

> The idea that a person is at fault when something goes wrong is deeply entrenched in society. That's why we blame others and even ourselves. . . but in my experience, human error usually is a result of poor design: it should be called system error. Humans err continually; it is an intrinsic part of our nature.[4]

Human-centered design requires upper-level management to trust the designers to do the research and find those system errors.

As I said before, UX design doesn't start with answers—it starts by looking for problems. Otherwise, following our Henry Ford example, we'd all still be riding around on horses.

CHAPTER 4

Analytical Leadership and Human Metrics: You Are What You Measure

> *Think before you speak.*
> *Read before you think.*
>
> *—Fran Leibowitz*[1]

I've worked for companies that had teams dedicated to research and data analysis. Some companies even had entire departments but, for some reason, the UX team never collaborated with the analytics department. There was a whole team of researchers and a wealth of information, but the reports they made were only for the executives.

If we were lucky enough to sit in on a meeting, we would catch parts verbally reported to management. Data was being hoarded between departments and leadership, but unfortunately, this seems to be the norm. UX teams are left to create their own data, pulling from their own research.

On one occasion, I needed to make a dashboard, and when I approached one of the data analysts, he was incredibly concerned.

"I do the dashboards," he told me.

"Yeah, I know. That's why they sent me to you," I responded, confused. The dashboard I was trying to make required some very simple metrics. The data analyst's dashboards were much more complex. Instead of having a collaborative attitude and helping me with the small amount of data I needed, the analyst immediately became defensive. It was almost like he thought I aimed to take his job.

Stakeholders should be able to trust designers to use data because it's so integral to the design process. Data and metrics determine goals and measure success. Leadership already knows this, but the metrics they are focused on the most are financial. Finance departments hold large amounts of data, but that information is off-limits to the rest of the business. For designers, seeing where customers are spending their money is incredible insight. These metrics tell us what users value, but because it is held so close to management's chest, it doesn't do us any good.

Every company has the ability to integrate data into their business, and they can start right now. First, audit the data you already have, and work from the inside out. With an idea of the data you hold, you can assess where you stand as a company. You would be surprised by what insights are already in your possession if only you looked through that data.

Then, initiate a self-analysis, which requires impartial data analysts. It's very easy to present data in a way that makes things look more favorable than they really are; I see marketing research used this way all the time. Having the most objective data allows you to see situations as clearly as possible.

That data needs to be seen by the rest of the teams. Poor communication within companies limits designers' abilities. It doesn't matter how much data you have, if there are barriers in place keeping collaboration from happening, the product and the users suffer. Data should be accessible to all who need it and responsibly and securely handled.

Leadership should take on a more analytical response and embrace the opportunities data has to offer. Data helps to eliminate those gut instincts HiPPOS tend to have. Just by choosing to use metrics beyond finances, stakeholders have the ability to make more complex data-driven or data-informed decisions.

Data-driven decisions are directly influenced by the numbers; you're going to do what they tell you to do. Being data-driven has its limitations since metrics don't always reveal the entire picture. The scope is important when letting analytics lead a decision, so it might be best to have data make more simple A/B choices. If you're data-driven, you're looking for validation.

Data-informed decisions, on the other hand, give context and allow for critical thinking. All of the user research and experiences paired with analytics are considered, but the human element is still behind the actual decision made. Typically, businesses have shifted away from data-driven strategies and moved towards data-informed strategies as pure numbers can't lead a business. Here, data acts as a check to that gut instinct you may have. Data from research doesn't threaten stakeholders' authority; it strengthens it. If you're data-informed, you're exploring new options.

Gathering usable data requires impartial research and a systematic approach to how real users interact with a product. Leadership often finds research to be expensive and a waste of time, but the UX team is most likely doing it anyway. Designers can't move forward until they have a better understanding of the users and their habits. Don't expect lab coats and clipboards. It may be as simple as seeing a few users actually interacting with the product or even having a conversation with a couple of users. The data has to be objective for it to be useful, but once collected, it can be applied in a multitude of ways.

One of the greatest tools UX designers have is personas. Personas are research-based profiles of typical users, fabricated from fictional names, ages, backgrounds, and photos. These fake people are reference points for researchers and stakeholders to keep user needs in mind. Personas offer an empathetic view into users but, again, they don't stand in for the real deal. Additional research and conversations are necessary for the best product possible.

Analytical leadership does not mean you have one arm tied behind your back by data analytics. It has bettered so many products that I've helped develop. While working on a program for a medical call center, a data analyst approached me with a solution to a problem we identified. The incoming calls from patients would queue up, and then the user (the call center agent) would select their next call. The problem was that the user could see the list of patients' information (insurance, reported problem, type of patient, etc.), so agents were cherry-picking the easier and shorter calls. The data analyst's solution was simply to click one button to receive whatever the next call in the queue was.

At first, I was totally against it. I had never seen this software pattern before, and to me, it seemed counterintuitive. I swallowed my pride though and decided to check it out and test it. To my surprise, the solution was extremely well received by almost everyone in the call center. The data didn't lie. It allowed agents to focus on one task at a time rather than parsing through every potential call. It simplified their workflow, and it solved the cherry-picking problem.

A closed mind is not human-centric, and when data is presented to you, you can't ignore it. If I had ignored the data analyst, who even had prior experience in a call center,

I would have strayed farther and farther away from the solution. Information needs to be efficiently shared between users, designers, and stakeholders. A more human-centric integration of metrics in UX could help redefine key business strategy, solving core issues and creating the most promising opportunities for a return on investment.

If you only respond to KPIs such as revenue and profit, you're limiting the potential for real growth and a broader vision. There is a range of metrics that businesses can use to guide their progress.

A North Star metric provides an overview of a product's overall status. As a product changes, it responds accordingly. This metric doesn't change often, so when it does, it's a good indicator to see if you are on course for those annual target goals.

A proxy metric can be used to stand in for immeasurable variables. Proxy metrics change more frequently and are easy to optimize for taking the place of unobservable data—leadership performance, value delivered, etc.

Counter metrics are paired with key metrics to measure necessary adjustments for balancing. They act as guards during optimization, so when your North Star metric shifts, you aren't negatively impacting other parts of your business.

Lastly are **leading and lagging indicators**. These metrics measure current performance and help stakeholders determine current and future actions. Leading indicators are *benchmarks* to reach goals. Lagging indicators *track* the completion of these goals, showing current performance and standing.

None of these metrics are useful unless they are paired with a human-centric approach. Accepting data in an objective and scientific manner doesn't just benefit profit, but also improves the well-being of users, customers, and employees. With a humble and honest approach, productivity and performance are sure to increase. It takes leadership and management to incorporate these values from the top down.[2]

With data applications that are paired with AI and computer learning, analytics offer descriptive, predictive, and prescriptive advancements. Data can help define a business strategy or save money by finding problems before they happen. Risk is reduced when implementation is data-driven. Complex systems can be simplified for more desirable customer journeys. These ideal situations are only possible when leadership trusts the numbers *and* their designers.

A survey shows 67 percent of senior managers or higher were not comfortable accessing or using data from their tools and resources. Even with strong data-driven cultures, 37 percent of those who responded expressed discomfort. The numbers don't lie; data should be used by everybody. 57 percent of companies where only key team members were responsible for analytics exceeded their business goals. In contrast, 82 percent of companies where *all* employees were responsible exceeded business goals.[3]

The human-centered approach to design and business inspires leadership to bring the best product to the user through the best experience for the employees. A HiPPO or a ZEBRA could derail a project when neglecting the necessary research and analytics for development, but when humanity is retained, everyone wins.

CHAPTER 5

Can You Trust Your Team?

> *A team is not a group of people who work together. It is a group of people who trust each other.*
>
> —*Simon Sinek*[1]

For every business that doesn't have a UX department or team yet, I have to warn you: the UX savior does not exist. Hiring a single UX designer with the expectation that they'll be able to fix all of your problems is a mistake. It takes a team, not a singular designer, to execute a delight-level design.

I've been in too many positions where I was hired with the expectation that I alone, as a UX savior, could save a project or build a product. Buzzwords like "ninja" designers or "rockstar" developers are too often used to advertise for positions in the tech industry. This only widens the gap between designers and stakeholders as the expectations for shiny and innovative design increase while the workload for a single designer (whether they can handle it or not) has its value diminished.

It doesn't help that solo designers, those without a UX department, answer to other department heads: the chief information officer, the VP of technology, or a marketing director. User experience should have its own department and vertical integration instead of random organization. By understanding what goes into UX, stakeholders can be more tactical in their UX team building.

Truthfully, UX designers should be led by a head of design. This team leader should be providing an overall vision for everything that a brand, service, and product touches, evangelizing on design and its importance. To create a strong UX presence in a company, there needs to be a cross-functional design department. The vertical organization of the department should have designers reporting to a director or manager of design. This manager would then report to the head of design. Running horizontally to this hierarchy, the UX design teams would each have their point of contact with finance, human resources, marketing, and other departments. Cross-functional teams are necessary for modern organizations to create human-centered products for employees and customers alike.

On a day-to-day basis, designers should be led by more mature designers, people who understand design thinking. Putting UX designers under an engineer or a marketing manager may work to an extent but, with an actual designer, they'll foster a culture of learning, promoting continuous improvement within the UX department.

If you do have only a single UX designer though, there are still ways to start building that necessary culture. Your designer should be present in every meeting that concerns their project; second-hand information and requests don't give designers the full picture. They need to hear directly from the boss, the clients, or the customers in order to develop design briefs.

The design brief is a document that outlines the core details and expectations of your project. A good design brief will set the tone by clearly stating the goals, expected quality of outcomes, and project deliverables. As the solo UX designer, I did occasionally gain permission to attend those meetings, but I was told not to ask any questions. What should have been a collaborative, exploratory discovery session ended up being awkward and tense. How does that get everyone on the same page?

Design briefs establish the shared vision of the challenges ahead and possible solutions for all members of a team. You need designers to make these documents not only for themselves but for non-designer management as well. These design briefs are meant to be shared with stakeholders and management; they synchronize the expectations of everyone involved. Again, if you only employ a single UX designer, letting them be the one to build that design brief—instead of a marketing manager looking over their shoulder and giving instructions—will result in a better product and experience.

It takes that little bit of initial trust to build a real relationship with designers. One of the issues I've noticed over the years is how stakeholders' lack of trust can negatively affect the team and product. Business educator Stephen Covey put it best: "When trust is high, the dividend you receive is like a performance multiplier, elevating and improving every dimension of the organization."[2] Trust goes both ways; it's a product of the relationship.

Uniting behind a common goal and a continuous improvement mindset is the key to fortifying that relationship. With clear milestones, everyone understands expectations and remains on the same page. Managers and stakeholders should trust that the designers will execute the agreed-upon plans. If that trust breaks down, there's a tendency to become frustrated. That's why shareable assets like design briefs are so important to uniting a team; they create a common jumping-off point for creative communication.

Clear communication is necessary to build on complex and innovative design ideas. Designers have a range of personalities, and oftentimes, their social levels differ from that of traditional leadership. Creatives can be shy and non-confrontational, and more assertive methods of communication might sour relationships. Persistence and patience can pierce through the communication breakdown, however. As long as stakeholders respect the established boundaries of roles and are self-aware enough not to cross them, trust and healthy relationships can form.

Again, it takes time to build trust. If your company has high turnover rates with UX contractors, then there is no foundation to build upon. Contractors are not invested in a company or the technology they're building long-term. They will often not write code or build proper designs because the intellectual property will then belong to the company. Management should be invested in their designers as well as the product.

Firing a contractor just to hire someone more junior may give management short-term control, but this robs the UX team of competent designers. It creates an ever-turning revolving door, making the design process even more difficult. Incompetent people end up hiring more incompetent people. Experience isn't everything, but if you're hiring someone on the "greener" side to have less pushback, your end product is bound to suffer.

If you are looking to build a UX team, you are better off hiring at least four generalists. These team members tackle all aspects of a company's UX needs: information architecture, user interface, human-computer interaction, research, etc. They should also have a level of business savvy to complement your company's business goals. During that interview process, you'll be able to pick up on what their strengths are; here is where you'll hone in on balancing your team. Personally, my strengths lie in information architecture and user interface. I can design clean, intuitive interfaces that simplify complex information and interactions. You wouldn't want to hire *another* generalist who is adept at user interface. Build a skillful team that will support each member and, in turn, your business.

If you don't know what you're looking for, an even better decision would be to hire a director or executive to lead the UX, along with a hiring budget. The only way this will work, though, is if management trusts the UX executive to make the right choices during the team-building process. Your UX executive should be on board with a human-centered culture. If not, the company and department culture will suffer, and so will your UX design. It pays to have competent leadership; otherwise, employee engagement will decline, and you will be more likely to lose new hires.

Human-centered management doesn't forget that employees are human, just like you. Respect the UX designers and their process, and the best product will be created for the users. Lose that respect, and you're 26 percent more likely to lose employees.[3] It takes self-awareness and humility to recognize toxic culture, but if that toxicity grows out of control, top talent will keep their distance from your company.

What it comes down to is: if you can't trust the UX design team you hired, who can you trust? You hired them, and they are there to perform a job you specifically cannot do. The UX designers, like users, will take time to build up loyalty to you and your company. Management cannot afford to assume they'll be immediately trusted, and when that trust is broken, it often never heals.

In reality, all it takes is one good leader to keep a team together and focused on the goals at hand. I was once in a leadership position where two of my subordinates often told me, "The only reason I'm here is because of you." What they appreciated the most was my ability to lead and mentor. That doesn't mean I told them what to do in just the right way. I created a space that was conducive to their growth. When I left the company, they both ended up leaving two months later. They no longer had anyone in their corner.

A happy, well-oiled UX team is a special thing. There's no drama between coworkers or power-hungry leadership. It's an overall positive environment for success. Communication easily flows between teams, and you don't have to worry about choosing the right words to tiptoe around stakeholders. I once had an impulsive manager who was easily provoked. Even though no one ever spoke ill of him and team members only wished to discuss the work at hand, he would take every poorly worded opinion personally. No one wants to share their thoughts with someone who will jump down their throat. By being transparent and open, everyone will be more willing to share ideas, providing a wider variety of perspectives.

Realistic expectations and a company culture built on trust are important for building and managing user experience teams. UX departments should be integrated throughout a company and focused on continuous improvement for all other departments and stakeholders. As clearly shared goals are reached, revenue and customer satisfaction are bound to grow. If you pin your hopes and dreams on a UX savior, you're bound to be disappointed across the board.

CHAPTER 6

Omnichannel Experiences: Thinking Beyond Screens

> *Your brand is a story unfolding across all customer touch points.*
>
> —Jonah Sachs[1]

When was the last time you bought a plane ticket? Say you sit down at your computer and open your email inbox, only to see a discounted ticket price in an email from the airline. You click on a link, are taken to the airline's website, and start to fill out your information. You realize you need to head to the store to pick up milk, so while you're on the bus, you finish paying for the ticket on the airline's app on your phone. When you're finished, you receive a QR code that you'll be able to use as your ticket when you leave for your trip. You also see options to purchase trip insurance and book a rental car. By the time you arrive home from the store, you have your itinerary and a receipt from the airline in your email inbox.

The entire interaction you've had with the airline, from buying your ticket to reading that final email, can be classified as the airline's omnichannel experience. All of that information can be easily pulled up from any of their channels, even by the flight attendants at your flight's gate. Omnichannel experiences offer seamless interactions across the digital world and the physical world.

Omnichannel refers to the user interacting with companies and organizations across multiple platforms, devices, channels, and applications. A customer's journey to completing a task can start in the physical world and end online. It may move from a phone call to an Internet browser to a brick-and-mortar store. The key to a delightful omnichannel experience, however, is the effortless and seamless transition between channels. Multichannel experiences also work across more than one channel (i.e. online, phone, and email), but there are perceptible gaps in service delivery. A company's services might be accessible on both a desktop browser and an app, but the app doesn't offer the full experience offered in the browser. The company's system doesn't communicate well between channels, leading to a disappointing experience.

In order for all channels to communicate properly, management must adjust their strategies to promote a seamless experience. This requires organizational leadership to keep teams, from marketing to engineers, on the same page—everyone has to be working towards the same human-centered goal.

Once the team is unified, they'll be in a better position to address real problems and opportunities for improvements. With time, a consistent experience will develop for leadership, designers, and, in turn, users.

The Customer Journey

Even with the goal of an omnichannel experience, every product usually has its own key base experience. For example, Instagram works best as a phone app. Adobe Photoshop is most useful when opened as a computer program. Despite this, you can access each product via other channels. Instagram can be accessed in a web browser. Photoshop has an iPad app. Therefore, the base experience is dependent on what the users want and need.

It comes back to conducting any necessary research, talking to the users, and finding out how they are using your product (or the context of their use). Users might be gravitating towards using a phone or their desktop. By determining the context, you can understand what aspects to continuously improve upon, working towards that ultimate goal of a delightful UX. Consistency is essential here; achieving your goals means ensuring users have the same (or at least very similar) experience across all channels.

Sometimes, the customer journey is purposely limited to keep users on the platform that most benefits stakeholders. For example, Instagram is purely optimized for phone applications. Trying to use Instagram on a desktop browser or even a tablet is frustrating: photos remain compressed, UI continues to be clunky, and the ability to upload content is extremely limited. Of course, some apps are device specific, but with a parent company such as Meta, they absolutely have the ability to bring Instagram to an iPad. The lack of device optimization most likely comes down to a conscious business decision—it's more efficient to monetize the mobile experience by selling advertisements and recovering user data than host additional platforms. In this instance, the customer journey suffers because of business limitations.

A customer's journey should be enjoyable, no matter the channel. It doesn't matter if they mainly use a tablet or a phone for your product. If they start filling out a form on their tablet and then decide to switch to their phone halfway through the process, they should be able to pick up right where they left off. This requires advanced levels of connectivity between systems, but this is what users are expecting nowadays.

Not only is the experience expected to be seamless, but it's expected to be personalized to the user. This begins with the initial offer: if I prefer using email, your company will reach out to me via email. That offer should route them to the part of the omnichannel experience that they will enjoy the most. You need to understand the user—their preferences, their behaviors, the right tendencies—and that takes time, research, and plenty of data.

A data lake, or a centralized repository for storing and securing large amounts of data, is necessary for each channel to access the same information. Whether it be for a phone or a website, each one has its own layer and pathway to the data lake; the shared repository means that the data is the same between platforms. If you update the data for the lake, it's updated for all channels. The consistency of data is a foundation for positive omnichannel experiences. When designing customer journeys, every team involved should have access to the same data to ensure proper cross-functionality.

Team Management

As I've said before, there is no UX savior—you need a diverse team that can handle the necessary aspects and platforms of your product. You cannot expect your design team to inherently know how to design an expansive, cross-functional experience without the right toolset and proper management.

Cross-functional teams need to be unified with a North Star objective for the product as well as more personalized goals for their departments. Your marketing team should be aware of the goals set for your web development team and vice versa. Leadership should enable the entire company with a long-term road map that unites every team under one roof.

This alignment is what defines an omnichannel experience: a seamless experience no matter how the user interacts with your company. Every little objective should point to the overarching vision for the company. Transparency between teams should exist in order to achieve these goals.

This level of open communication is the cornerstone of a cross-functional team: tear down the walls between teams and allow for super transparent over-communication. Leadership should encourage horizontal communication between departments and vertical communication with stakeholders and leadership. The mentality of promoting competition between departments is now an antiquated method of encouraging productivity. Modern businesses cannot thrive, let alone survive, without free and open communication.

If your internal organization requires an overhaul to reach optimization for omnichannel experiences, do not put that burden entirely on management. All parties should be involved in creating a more open structure. Otherwise, the company will slip back into old ways. McKinsey and Company consulting firm puts it best:

> Management sets the goals and overall direction but doesn't simply hand employees a new way of doing things. Instead, employees help build the new system, and that ownership helps implementation engage initially, then stick for the long term... In fact, our experience has found that implementations led by the workforce are five times more sustainable than those led by management. Organizations that embark on successful cross-operational transformations keep this in the front of their corporate minds.[2]

Siloed teams will not build the best product, especially when modern products are expected to work on every platform and channel. The results will turn out choppy and nonuniform. When a user comes to your product, they should be delighted, no matter if they are in your store, on your website, using your app, or whatever it may be.

When It Works

When I recently sold my car, I did not expect to have a positive omnichannel experience during the process. I sold my car to the online used car dealer, Carvana, after comparing offers with dealerships and other websites. With Carvana, the entire process occurs through your phone.

I walked around my car, taking pictures from all angles and snapping a shot of the license plate. They gave me an offer based on the information I provided, trusting the user in this instance. I agreed to the pretty generous price and scheduled an appointment for pickup.

One of their flatbed trucks came to my house, and Carvana employees inspected my car. They signed off on the inspection, loaded up my car on the truck, and waved goodbye. Three days later, I received a deposit in my bank account.

Overall, this was an extremely easy omnichannel experience that resulted in a positive customer journey. From the simplicity of using my phone to take pictures, to coordinating the inspection and pickup, to even the money transaction, they nailed all channels. Had I gone to a dealership, this could have been a frustrating experience of haggling with a representative for hours over price before driving to another dealership for the chance at a better deal.

Instead, it was a breeze because Carvana's systems can efficiently communicate between the company's mobile app, website, and pickups. Carvana trusts the user to facilitate most of this process, which is a huge risk on their part but results in fewer roadblocks and hiccups.

When It Doesn't

For 10 years, I've worked with electronic health records in healthcare, and I can honestly say: these terrible systems are far from being omnichannel experiences. They are barely even multichannel experiences. Electronic medical records are incredibly important for healthcare providers and insurance providers, but these systems are Frankensteined together; they are incapable of communicating between companies, providers, or platforms.

One of the biggest problems with electronic health records (EHRs) is the inability to transfer records between healthcare providers. If you go to one provider, they will have their own data set. Then, if you go to another, they will have a completely different data set. On top of that, so many healthcare providers still work with paper records. You can't have patients simply transfer their information. Each time they see a different provider, they have to input all of their information from scratch.

These records are hard to access even within their own systems. Almost no EHR is mobile-optimized, so checking the information on your phone is impossible. In addition, one EHR requires a Windows operating system, so if you're on a Mac, you have to run a software called Parallels to run Windows on your Mac. In some instances, even if you fill out your information online before visiting the provider, you'll still have to fill it out on paper forms when you arrive for your appointment. It's a broken system.

The lack of compatibility between EHRs and the inability to have that data shareable across the cloud for healthcare providers further bogs down the healthcare system. You know who suffers because of this? Patients.

If your primary care provider prescribes you one medication but your specialist prescribes you another, that information might not be shared between the two doctors. Those two medications, if not checked, could lead to a potentially fatal interaction. When EHRs are unified between providers and easy to access across all channels, the number of preventable deaths could be reduced.

Fortunately, current initiatives, such as with providers and Google, do plan to consolidate all of these records. The company's massive infrastructure and finances mean it can carry out such a large-scale, significant project. There is a downside to having a large tech company sort health records, though Decisions would be made to seize any opportunity to make more money rather than promote healthcare as a given human right.

Users demand seamless experiences across all channels, including websites, mobile apps, social media, and in-person interactions. The benefits of promoting omnichannel experiences will only increase your users' enjoyment and your business's revenue. With more open team management and accessible data, the limit of how far your product can reach is up to you.

CHAPTER 7

Continuous Improvement

> *From error to error, one discovers the entire truth.*
>
> —*Sigmund Freud*[1]

The amount of computational power in our pockets is a testament to continuous improvement. When computers used punch cards to transfer data, holding only about 80 bytes, it would have been inconceivable to think about the amount of data circulated almost instantaneously today, yet here we are. And these numbers will only grow to more inconceivable sizes as technology advances.

The momentum of technology will never slow to a halt; it will continue to evolve. If you are able to design with this in mind, you can tackle problems of any scale. It is the nature of continuous improvement. It's never stagnant, always innovating.

For me, it's a personal mindset. I always refer to the phrase, "Good enough is just good enough." I can't do "just good enough." I'm never truly happy with what I deliver; sometimes, I will over-fixate on a product's flaws. That isn't to say that the final product isn't good. Products have delivery dates, and perfection is impossible. My expectations are incredibly high, but an iterative process is required to actually reach them. I've learned to break down the steps necessary to reach those big-picture goals. It's the ability to zoom out, see the forest for the trees, and then zoom back in.

So, what are you looking for in the forest? Initially, it is the minimal viable product (MVP). With an MVP, that iteration of the product is a checkpoint: the minimum effort needed to evaluate the product and have a sense of direction for the project. Your MVP will never be your final product, but it is the start of a greater journey. Be mindful during the design process and acknowledge that each product iteration is a touchstone on the way to larger goals. Even then, your goals will grow with your business.

Goals need to be set with both designers and stakeholders in the room. Teams are kept separate by a self-imposed hierarchy, which only constrains productivity and creativity. Strategy meetings are held between VPs and other higher-ups, where they set goals before even consulting the design teams. I've often found that these goals are established by solely considering financial data. The design team is left asking, "Where's the research? Where's the market validation for this feature?" Company culture is tested in these moments as you learn if your leadership is a HiPPO or willing to listen to feedback.

A singular vision or goal is necessary to engage your employees. That means engaging them when those goals are set. If you believe in people, they'll believe in your mission and gain fulfillment from their job. This human-centered mindset establishes a goal for employees to rally around. With a sense of purpose, employees can focus on continuous delivery.

Again, human-centered design has to be integrated into the company culture at all levels. Teams need to be built with a mix of doers and dreamers. They can't all be one or the other. With too many dreamers, nothing will get done. With too many doers, nothing will be long-term. Leadership should be able to inspire and execute plans for each product iteration. Everyone needs to be mindful of the team as well as the user—self-awareness is necessary for goal setting. Once those goals are agreed upon, steps need to be taken to reach them.

Agile should be seen as a principle and mindset that focuses the roadmap for incremental improvements. Yes, be ready to deliver as quickly as you can, but recognize it's just one iteration of many. You should always strive for a better experience, a better situation, or a better product. Every design impacts someone; design intervenes systems and interrupts lives. If done responsibly, that positive change can become a positive trend.

Iterations require recognition of errors in order to improve. A humble nature is required. There will never be a "one-and-done" situation; you can't rest on your laurels after a single iterative success. Existing problems and pain points should be identified so that design thinking and the design process can better them. By the end, a more ideal experience is realized.

Ignoring the iterative process would be foolish. It'd be like trying to climb a flight of stairs in a single leap; it just won't happen. You need to take multiple steps, and when you get to the top, there will be another flight. There's always another flight. If continuous improvement is not adopted as a mindset, you'll be left standing on the stairs with everyone passing you on their way up.

Technology is the biggest enabler in the modern world. It lifts people from stagnant positions by giving them abilities that weren't previously available. With a goal to leave the world better than you found it, continuous assessments should be made to understand the true roots of problems and how to solve them.

You learn very quickly: there's always more than one way to solve a problem. Continuous innovation, in small increments, allows for more prototypes to be developed, which then turn into more products to deliver. This continuous stream of updates and products provides more opportunities to keep customers and users more satisfied. Customer experience will only improve as the products continue to deliver on their promises and goals.

"Good enough" will stall your business. Unless you're looking to design products that improve with iterations, expect to be surpassed by competitors. This means a focus on the user experience. Not every goal will be reached initially, but that's okay. Accepting a pivot or shift in direction does not signify a failure. Stopping improvement, either from disappointment or complacency, *will* turn into a failure; it will not stop competitors and the industry from passing you.

It comes down to stakeholders maintaining a conscious attitude during the design process. We have to think like film critics with self-awareness in relation to the medium. What does a critic want more than anything? Better art. Their intent is to elevate movies as a whole with an objective approach to finding what works and what doesn't. If you are looking at a product and trying to give feedback, there's a tendency to point fingers at someone and find out who is responsible for disappointing work. It's equally easy to look at a movie, point at a name on the screen—the director, the cinematographer, or an actor—and say they were the one who made it bad. The entire system needs to be recognized as a whole.

This includes acknowledging what *is* working. Solely cracking the whip to maintain momentum and continuous improvement is counterproductive. You need to acknowledge the work that people have done, give them a sense of ownership over the product, and unite everybody around your main objective.

At a previous job, two senior-level "negative Nancies" spoke with such conviction that their criticisms would suck the air out of the room. The team would proudly begin presenting a demo to them, but all of a sudden, the Nancies' negative comments would leave the team uninspired. I would have to take time after these meetings to really encourage the team to see the positive side: look at all these checkpoints that we crossed, look at how much closer we are to the end goal, etc. Without a positive check-in, how will designers know what is working not only in the product but also within your organization's structure?

Staying human-centered creates the best environment for maintaining success across the product and the team, and you'll need it. Market expectations are higher than ever, thanks to giants like Google. When Gmail came out, it performed one core function: it sent emails. Now, it can integrate calendars, data management, and third-party apps. So, if you're trying to come to market with a new email product, your MVP needs to be much, much bigger to be even remotely competitive.

This is sometimes used as a cop-out for delivering incomplete and less-than-useful experiences. Even I have been guilty of this. If the user or a customer is asking for something that is not really in your roadmap and you choose to put it off for later iterations, there will be consequences. Users and customers don't want to feel ignored. This is why strategy and goal setting is so important, which require data collection and metric analysis. KPIs and OKRs should be decided upon with designers, stakeholders, and users in mind. OKRs will set your goals, and KPIs will make sure you reach them. The metrics are only effective when both designers and leadership are aligned and following them.

A well-defined goal, the fortitude to reach it, and the choice to pick a new one—that's all continuous improvement is. Those goals should be chosen with a sense of responsibility and altruism. Design has the potential to change lives, and it requires the work of many. By unifying efforts under a human-centric mentality, we create opportunities to make the world a better place, again and again.

CONCLUSION

> *To find ideas, find problems. To find problems, talk to people.*
>
> *—Julie Zhou*[1]

At one of my previous jobs, I created a camaraderie with a peer. I really admire his work as a front-end developer; he is that rare breed of person who can be given a design and build it exactly as it was intended, possibly even improving it. He recently reached out to me with freelance work on an app he was developing. He had already started on it and even had investors, but he needed someone to work the UX for it.

When I initially looked at the app, I saw it would be competing with huge, established giants. I asked my friend to talk me through his idea and why he believed it could make a dent in the market. He didn't have an answer. He had already begun working on this app idea and already had a design. He had made a logo and chosen a name. He didn't have any market research though.

I saw that I needed to go way, way, way back instead of working on the UX or UI. I determined the product-market fit and built a presentation with 50 slides to show him that this app wouldn't be fruitful. I laid out all of the data, insight, and research I could get my hands on. He was blown away, admitting he hadn't thought about the app within the greater scope of the market.

Even people who *have* extensive knowledge about user experience can be stumped by the design experience. This app is an example where investors were just being brought on board, so there was still room to pivot, but this isn't always the case. Some products end up on the market for years before a UX designer is called on to work on them. An inflexibility sets in, making it hard to do anything but the same thing over and over again.

If your technology is incredibly innovative and radically disruptive, your product may stand out among the rest of the market, but we all know how rare that actually is. With research and competitive analysis, you stand a chance with an iterative approach to improving your product. Unless the heart of the technology is identified and human-centered design is in place, the poor UX designer you hire will just be putting lipstick on a pig.

What I'm promoting is a shift in culture, and hopefully, it catches on quickly. Human-centered design isn't a trend to be discarded in the coming years. It's a new foundation for how technology will be defined in a world still just burgeoning from global connectivity across the Internet. The ability to communicate and share ideas within a worldwide community requires a standard of addressing each user as a human with certain unalienable rights.

Right now, poor business methods and leadership have held designers back, but they are an untapped resource for rapid innovation and implementation. By trusting your designers, you create an environment and a product that progressively builds on itself. Does this mean that a business sensibility isn't necessary? No, of course not; business sensibility *is* good design. Just like any business, research, and planning are necessary for consistent growth. Design may not move from quarter to quarter, but continuous improvement isn't tied down to the whims of the market.

A conscious effort by leadership to implement a human-centered culture within an organization will lead to happier users. You can start now by explicitly promoting these ideas in all internal and external communications: users and employees should know that they are seen and heard under the standard that they are fellow human beings. When an organization strays from that human-centered standard, it will be given feedback in response. The more this spreads across industries and becomes baked into global culture, the more delightful experiences will become common across our daily lives.

As someone who has extensively worked with healthcare organizations, you would be surprised to know how often even patient-centered design is ignored. The more people are pushed away from the very designs they are expected to use, the faster they will be drawn to true human-centered experiences. If you don't want to be left behind, it's time to earnestly talk with your design teams about making significant changes.

What may seem like an intimidating conversation is really the beginning of an exciting journey. There are so many paths forward that you and your organization can take, but the most fruitful ones include real discussions with UX designers and users. Management styles of the past have their limitations; it's with the tools and understandings laid out in this book that I hope you see that trusting your UX teams will lead to unseen success. UX design is truly in its fledgling years—it will only continue to grow alongside the world's technological advances.

ACKNOWLEDGMENTS

Many thanks to Saribel and José for making my time at Sabiamed so much easier and more fun. It was my first experience in UX, and it informed how I approach design today.

It may have been a short time together with “el griego loco,” Erleh, at Westgate Resorts, but I greatly appreciate how he made the workplace enjoyable while teaching me a great deal.

Swati’s influence during my time at Mindtree cannot be understated. I can’t thank Swati enough for believing in me from the start and instilling so much confidence.

Thank you, Rafael Padilla, for all the stories; I learned a great deal from you and our conversations while at TravelClick. I absorbed all the wisdom that I could.

Cato, Chris, and José, thanks for making the team at BioPlus so special. It was great working with all of you.

Thanks to Scott and Arjan at BrightRay Publishing.

REFERENCES

Introduction

1. Herbert A. Simon, *The Sciences of the Artificial*, 3rd ed. (Cambridge: The MIT Press, 1996).

Chapter 1

1. Drew Hansen, "Myth Busted: Steve Jobs Did Listen To Customers," December 19, 2013, https://www.forbes.com/sites/drewhansen/2013/12/19/myth-busted-steve-jobs-did-listen-to-customers/?sh=e087e4087f32.
2. Don Norman, "Systems Thinking: A Product Is More Than the Product," December 3, 2018, https://jnd.org/systems-thinking-a-product-is-more-than-the-product/.
3. Phillip Britt, "Customer-Obsessed Companies Pull in 2.5 Times More Revenue," September 7, 2021, https://www.destinationcrm.com/Articles/CRM-Insights/Insight/Customer-Obsessed-Companies-Pull-in-2.5-Times-More-Revenue-148814.aspx.
4. Jacquelyn Bulao, "36 Astonishing Customer Experience Statistics for 2022," Techjury (blog), September 5, 2022, https://techjury.net/blog/customer-experience-statistics/#gref.
5. "Design Thinking Defined," IDEO Design Thinking, accessed September 29, 2022, https://designthinking.ideo.com/.

6. James Womack and Dan Jones, "Lean Thinking at 20: A Q&A with Jim Womack and Dan Jones," September 28, 2016, https://www.lean.org/the-lean-post/articles/lean-thinking-at-20-a-qa-with-jim-womack-and-dan-jones/.
7. Robert L. Read, "Human Centered Design vs. Agile Development: Adaptive Balance," March 4, 2017, https://medium.com/agile-government-leadership/human-centered-design-vs-agile-development-adaptive-balance-8fd51e652e4f.

Chapter 2

1. Hugh McFall, "Can design transform social impact?" January 10, 2016, https://medium.com/hugh-s-blog/designing-for-social-impact-838b0724c96c.
2. "A brief history of human-centered design," Queensland Government, accessed September 29, 2022, https://www.forgov.qld.gov.au/__data/assets/git_bridge/0022/174172/franchises/forgov/learning-modules/A_brief_history_of_human_centred_design/#/.
3. Leyla Acaroglu, "Tools for Systems Thinkers: The 6 Fundamental Concepts of Systems Thinking," September 7, 2017, https://medium.com/disruptive-design/tools-for-systems-thinkers-the-6-fundamental-concepts-of-systems-thinking-379cdac3dc6a.
4. Michael Goodman, "Systems Thinking: What, Why, When, Where, And How?" accessed October 3, 2022, https://thesystemsthinker.com/systems-thinking-what-why-when-where-and-how/.

5. "Prediction: The future of Customer Experience," McKinsey & Company, April 11, 2022, https://www.mckinsey.com/us/our-insights/prediction-the-future-of-customer-experience.

6. Jack Kelly, "A Hard-Hitting Investigative Report Into Amazon Shows That Workers' Needs Were Neglected In Favor Of Getting Goods Delivered Quickly," October 25, 2021, https://www.forbes.com/sites/jackkelly/2021/10/25/a-hard-hitting-investigative-report-into-amazon-shows-that-workers-needs-were-neglected-in-favor-of-getting-goods-delivered-quickly/?sh=7204806c51f5.

7. Annie Palmer, "Amazon warehouse workers suffer serious injuries at twice the rate of rivals, study finds," April 12, 2022, https://www.cnbc.com/2022/04/12/study-amazon-workers-suffer-serious-injuries-at-twice-rate-of-rivals.html.

8. Jason Del Ray, "Leaked Amazon memo warns the company is running out of people to hire," June 17, 2022, https://www.vox.com/recode/23170900/leaked-amazon-memo-warehouses-hiring-shortage.

9. Erika Hall, *Just Enough Research* (New York: A Book Apart, 2013).

Chapter 3

1. Patrick Vlaskovits, "Henry Ford, Innovation, and That 'Faster Horse' Quote," August 29, 2011, https://hbr.org/2011/08/henry-ford-never-said-the-fast.

2. Ward Cunningham, "Debt Metaphor," February 14, 2009, video, 4:43, https://youtu.be/pqeJFYwnkjE.

3. Michał Mazur, "What is design debt and why you should treat it seriously," August 13, 2018, https://uxdesign.cc/what-is-design-debt-and-why-you-should-treat-it-seriously-4366d33d3c89.
4. Don Norman, *The Design of Everyday Things: Revised and Expanded Edition* (New York: Basic Books, 2013).

Chapter 4

1. "Quote by Fran Lebowitz," Goodreads, accessed October 3, 2022, https://www.goodreads.com/quotes/56442-think-before-you-speak-read-before-you-think.
2. Y. Lee, C. M. Berry, and E. Gonzalez-Mulé, "The importance of being humble: A meta-analysis and incremental validity analysis of the relationship between honesty-humility and job performance," *Journal of Applied Psychology* 104, no. 12 (2019): 1535–1546, https://doi.org/10.1037/apl0000421.
3. Thomas H. Davenport, Nitin Mittal, and Irfan Saif, "What Separates Analytical Leaders From Laggards?" February 3, 2020, https://sloanreview.mit.edu/article/what-separates-analytical-leaders-from-laggards/.

Chapter 5

1. Simon Sinek, Twitter post, August 6, 2012, 12:18 p.m., https://twitter.com/simonsinek/status/232556392114974721
2. Stephen M. R. Covey, "The Business Case For Trust," accessed October 3, 2022, https://chiefexecutive.net/the-business-case-for-trust/.

3. Lory Li, "17 Surprising Statistics About Employee Retention," April 8, 2022, https://www.tinypulse.com/blog/17-surprising-statistics-about-employee-retention.

Chapter 6

1. Pranoy Kumar Maitra, "'Your brand is a story unfolding across all customers touch point' - Jonah Sachs," August 20, 2021, https://www.linkedin.com/pulse/pranoy-kumar-maitra-4e/.
2. Alejandro Sandoval et al., "Transform the whole business, not just parts," October 10, 2019, https://www.mckinsey.com/capabilities/operations/our-insights/transform-the-whole-business-not-just-the-parts.

Chapter 7

1. Norbert Juma, "Sigmund Freud Quotes From The Master Of Psychoanalysis," August 11, 2022, https://everydaypower.com/sigmund-freud-quotes/

Conclusion

1. Julie Zhuo, "Finding ideas," November 13, 2017, https://medium.com/the-year-of-the-looking-glass/finding-ideas-d9c07ca1a076.

ABOUT THE AUTHOR

Tommy Pagán's passion for design has always been accompanied by a strategic mindset. From a young age, he would immerse himself in doodling, envisioning innovative designs for various objects. This zeal led him to pursue a degree in graphic design, where he embarked on his professional journey as a user interface designer.

Tommy's deep dive into usability and accessibility during his early career exposed him to the life-changing potential of design. Tasked with creating touchscreen interfaces for emergency room self-registries, he witnessed firsthand how design can transform lives. This experience ignited his passion for user experience design, driving him to make a tangible impact through thoughtful and impactful design.

Motivated by an unyielding thirst for growth, Tommy relentlessly expanded his skill set, mastering visual design and product strategy. With an unwavering determination, he embraced user experience design, dedicating himself to shaping digital products, services, and businesses that truly enhance people's lives. Today, as a seasoned senior user experience and product designer, Tommy thrives on complex ERP and SaaS projects, utilizing his creative and strategic acumen to create profound impact for individuals and businesses alike. Tommy Pagán's unique blend of creativity and strategic thinking propels him to push the boundaries of design and leave an indelible mark on the world.

Tommy lives in Florida with his wife, Nilma, and their two kids, Gustavo and Penelope.